# BAMBOO OF JAPAN

Splendor in Four Seasons          Photographs by SHINJI TAKAMA

# BAMBOO OF JAPAN

## Splendor in Four Seasons

Photographs by SHINJI TAKAMA

Graphic-sha/Japan Publications

© 1986 Shinji Takama
Design & Layout: Kenichi Yanagawa

Published by Graphic-sha Publishing Co., Ltd.
1-9-12 Kudankita, Chiyoda-ku, Tokyo, Japan.

Distributors:
UNITED STATES: Kodansha International/USA, Ltd.,
through Harper & Row, Publishers, Inc., 10 East 53rd
Street, New York, N.Y. 10022. SOUTH AMERICA: Harper
& Row, Publishers, Inc., International Department.
CANADA: Fitzhenry & Whiteside Ltd., (195 Allstate Parkway,
Markham, Ontario L3R 4T8.) MEXICO AND CENTRAL
AMERICA: HARLA S.A. de C.V., Apartado 30-546,
Mexico 4, D.F. BRITISH ISLES: International Book
Distributors Ltd., 66 Wood Lane End, Hemel Hempstead,
Herts HP2 4RG. EUROPEAN CONTINENT: Fleetbooks,
S.A., C/o Feffer and Simons (Nederland) B.V., Rijnkade 170, 1382
GT Weesp, The Netherlands. Australia and New Zealand:
Bookwise International, 1 Jeanes Street, Beverley, South
Australia 5007. THE FAR EAST AND JAPAN: Japan
Publications Trading Co., 1-2-1, Sarugaku-cho, Chiyoda-ku,
Tokyo 101.

First Printing: May 1986
ISBN 0-87040-707-4

Printed in Japan

# CONTENT

# The Charm of Bamboo ——————————— Keiichi Ito

(Novelist & Poet)

I was born and raised in a rural area of the Ise District.  My father was a Buddhist priest and I grew up with the beauty and the sounds of the bamboo forest behind his mountain temple.  The scenery in that part of Japan is enriched by groves and thickets of bamboo which give it picturesqueness and unexcelled beauty.

Photographer Shinji Takama, the creator of this book, has been and will probably be forever attracted by the loveliness of bamboo.  I, too, have developed a lifelong love of the plants and they have become the subject of many of my poems and the background for scenes in my novels.

One of my poems is "The Song of the Bamboo."

Bamboos upon a hillside,
Drawing love from all the other trees
Bestow serenity upon the hill
And make it sometimes smile
Beneath a tender touch.
There is a coquetry so cool
My fingers almost feel it melt.

Close by that mountain stream
Where even in the summer warblers sing
I turn and leave the hill.
Only the bamboos bow to me.

Sometimes my days are spent on naught but gazing
At bamboos and listening to the warblers' trills,
And then I gather from the stream its sounds
And with them fill my empty creel.

To gaze at bamboos is to be gentled
And to yearn for union with the ones we love.
The bamboos stretch, they bend
And ever exercise.
I wonder
To what music do they sway ?

In this poem, I have tried to present a bright and positive thought about life.  That might be the influence of a sentimentalism bred by my boyhood among the bamboos, or perhaps, by some vague influence of my Japanese character which is inherently moderate and patient, nurtured as it is by the bamboo in the world around me.

Sometimes while riding through the countryside on a train, I will leisurely lean against the car window and spend a pleasant time gazing out at the bamboo–sprinkled scenery.  When I do so, I feel a desire to be absorbed by, to dissolve into the bamboos.  This, I think, reflects some sort of primitive religious feeling, the sense that man is an intimate part of nature.

6

When I was young, I spent some time as a soldier in northern China, and I recall one day walking through the yellow soil of the region's dreary mountain ranges and thinking of death. I felt that if I died, the earth would hold my corpse in silence and my soul would remain upon the surrounding trees and shrubs. That was my view of life and death. There was no resignation, but rather a bright affirmation fostered by the days I had spent in constant nearness to bamboo. That is still my view of nature and life.

Like those bamboo photographs of Takama's which I have seen on exhibition, the work in this book has a powerful appeal that recalls the whispering sound of bamboo leaves and makes it possible for me to feel again the breath, the very warmth of the bamboos' "skin." I have the sensation of strolling in some psychic world of bamboo. For both Shinji Takama and myself the bamboo is not a mere plant, it is a living being. Indeed, it is ourselves. In my poem "Scenery," I have written the following.

> A felled bamboo.
> Its spirited cry and final gesture
> Come before a brightening sky
> As those who share its fate
> Huddle together and smile,
> Strangely happy, waiting for their turn.
>
> From somewhere a clear laugh.
> No bird, it is a man who breaks the thicket's silence —
> Cutting
>
> The bamboos fall
> Sweeping through the Prussian blue in gentle arcs
> And softly, softly laughing.

In this poem I may say there is a kind of exchange of atmosphere, a slight one, but one in which are revealed the feelings of friendship shared by the poet and the bamboos. There you can see the world of beauty as poetry sketches the bamboo. However, the poet's view may be a little too abstract. In the case of photography, there is a more direct approach. The photographer's eyes bring man and bamboo together in mutual love. With his camera, man can truly embody the gentleness of bamboo, the loveliness, the strength and all the mysterious shapes and patterns. This he does with joy and a refreshing directness.

As I hold before me Shinji Takama's book of photographs, I feel a work bursting with happiness. And I sense within it a revelation of all the ways in which bamboo can bring serenity and relief to the soul of man.

## SPRING

Our universe is endlessly, constantly changing.

New things burst upon the scene, others disappear,

and I feel real joy at having the eyes and ears with which to observe this boundless providence of nature.

For me, spring makes its presence felt in the growth of the bamboos.

The land grows greener by the day⋯trees and grasses bloom one after another.

The plums and peaches reach the peak of their blossoming and with them the cherries

and camellias. It is truly spring and there is a lively mood upon the land.

Now, more than anything else, it is the bamboo shoots which speak to something deep inside of me.

They burst forth green and fresh and beyond all praise.

9

Bamboo and plum blossoms

10

Fresh signs of growth

Dancing leaves

Sprouting bamboo shoots

A new bamboo shoot

A bamboo garden

Twin sprouts

Tears

Magome spring

19

At the foot of Mt. Takachiho

Light and shadows

Bamboos and cherry blossoms

22

Camellia red — bamboo green

Bamboo roots and fallen camellias

Volcano seen through a bamboo grove

Vine on bamboo

26

Jikishi-an garden in Kyoto

Designs of nature

Twisted bamboo sheath

Shades of beauty

30

Stumps — aged and new

The red-tipped roots of a young bamboo

## SUMMER

I have always felt that life has meaning only when a man puts forth his best efforts,
when he lives it to the fullest.
If life resembles any season, that season is summer,
a time of passion and exuberance, a time when all the flowers
and grasses are resplendent in their blossoms and when the wild birds sing.
I love this vigorous stage in nature's life and take pleasure in the newborn bamboos
and the burgeoning of the lush green groves.

Fresh rain on a young bamboo

The elegant shoots of a "madake" bamboo

Crisp, pale beauty

Kyoto bamboo grove

On a rainy day

39

Weathering a typhoon

40

A sprinkling of yellow

Hillside bamboo grove

Wildflowers brightening a dark thicket

43

Summer  dream

Secret flowering

Whispering

46

Bamboo on a gorge

47

At Takeshima

Pink in a field of green and white

Black bamboo and wild iris

Twisted bamboo and orchids

Bamboo blossoms

## AUTUMN

With the onset of autumn, the colors of the hills and fields begin their annual change. Plants,
sensitive to temperature, anticipate winter's coming.
Colors change and forms are subtly altered.
Only the bamboo remains unchanged, calm and self-possessed.
And though the other plants shift to reds and yellows,
their brightness only serves to accentuate the beauty of bamboos.
In autumn, I go forth in search of contrasts and harmony of color,
and it is then that I discover the little things in nature that have so long gone unnoticed.

Patterns of light and color

Autumn

Residents of a bamboo thicket

The gentle song of bamboo leaves

Bamboos in ancient Nara

Persimmons in a bamboo grove

The reflected brilliance of blue sky

Bamboo and ivy

Late autumn

Striped bamboo and larches

Iso Garden

Stillness

66

Hydrangeas in bloom

Summer's verdure

The dignity of age

Accents of autumn

Odaigahara autumn

Sunset

## WINTER

As a child, I often played in bamboo thickets
and on snowy days I found great pleasure in kicking the plants' trunks
and watching miniature snowslides cascading down to earth.
Now, as a man, I wait for snow to come and beautify those same groves.
And I take pictures filled with severity and cheerfulness.
The sight of those sturdy bamboos and their pure white mantle is a deeply moving one that sweeps away my fatigue,
and I stand in awe of nature's shapes, designs and endless variety.

Snow-touched bamboo blades

Snowy patterns of light and shadow

Fleeting patches of leftover snow

Snow-girdled bamboo

Obeisance

Twilight

Snow blossoms

Radiant golden bamboo

Winter's designs

Solitude

Frozen beauty

Snow fantasy

Snowstorm

Swaying rhythms

88

Sunrise at Ibi

A hamlet in Wakasa

Bamboo grove at Yakushi temple

Yoshino in winter

Moonlit  night

## AFTERWORD

My home is the village of Ibi situated in the Nishino district, also called the province of Mino, northeast of Mt. Ibuki in Gifu prefecture. It is filled with the splendors of nature—majestic mountains, crystal-clear rivers and fresh, clean air. I was born and raised here, at the foot of the mountains of Ibi Castle.

Called into service during World War II, I bade farewell to my village and sailed off to Manchuria. At the end of the war I was detained in a prisoner of war camp in the Soviet Union. Three years of my youth were given over to that utterly dismal life, until at one time I had almost even given up hope of ever returning to Japan. Words cannot express the joy I felt when I was finally able to return to my village. The rivers and mountains I had yearned for and dreamed of were every bit as beautiful as I had remembered.

There are many bamboo groves around Ibi, and I became mesmerized by their beauty and engrossed with photographing their seasonal display—streaked with sun, wet with rain, bending with the wind and covered with snow. I remember especially the spellbinding view of the bamboo trees in snow. The pure beauty of the clean white snow looked as though God had spread a blanket over the land. This was a soul-stirring experience for me, perhaps because of the harsh life I had led as a prisoner of war. As I watched the bamboo trees withstand layer after layer of snow, I pictured my previous self in them.

93

It was my great good fortune to study with the renowned photographer, Ken Domon, but I continued working constantly with bamboo. I have done the different varieties as well as each flower, blade of grass, and tree growing in a grove. I have even gotten the nickname Takama, the bamboo specialist, and have devoted thirty years to photographing these trees. From concentrating first on the village of Ibi, my work fanned northward and southward, taking me throughout Japan.

Bamboo is truly the soul of the Japanese, and its affect on everyday life is beyond calculating. I believe that in no other part of the world is a material so entwined with a culture even as it changes with the four seasons. These seasonal changes are the focus of the eighty pieces of my work in this collection. I hope through this book that many others will be introduced to the enchantment I find in bamboo.

As a final word, I would like to express my thanks to Keiichi Ito for writing the introduction to this book, to the publisher, and to all who have worked on this in any way. I offer my deepest appreciation for their cooperation and assistance.

# Photographic data

## Spring

9   *Bamboo and Plum blossoms*
PENTAX 6×7
SMC TAKUMAR 300mm F4
f5.6 1/8 EKTACHROME

10   *Fresh signs of growth*
LINHOF SUPER TECHNIKA V4 4×5
TELE-ARTON 360mm F5.6
f16 1/30 FUJICHROME

11   *Dancing leaves*
LINHOF SUPER TECHNIKA V4 4×5
FUJINAR SC250mm F4.7
f11 1/125 EKTACHROME

12   *Sprouting bamboo shoots*
LINHOF SUPER TECHNIKA V4 4×5
FUJINAR SC250mm F4.7
f11 1/60 EKTACHROME

13   *A new bamboo shoot*
LINHOF SUPER TECHNIKA V4 4×5
TELE-ARTON 360mm F5.6
f8 1/10 FUJICHROME

14   *A bamboo garden*
⸱
15   LINHOF SUPER TECHNIKA V4 4×5
XENATAR 150mm F2.8
f11 1/60 FUJICHROME

16   *Twin sprouts*
LINHOF SUPER TECHNIKA V4 4×5
FUJINON SF250mm F5.6
f8 1/15 EKTACHROME

17   *Tears*
LINHOF SUPER TECHNIKA V4 4×5
TELE-ARTON 360mm F5.6
f11 1/15 EKTACHROME

18   *Magome spring*
PENTAX MX 35
SMC PENTAX M200M ZOOM35 F2.8 35-70mm
f16 1/250 FUJICHROME

19   *At the end of Mt. Takachiho*
LINHOF SUPER TECHNIKA V4 4×5
TELE-ARTON 360mm F5.6
f16 1/30 FUJICHROME

20   *Light and shadows*
LINHOF SUPER TECHNIKA V4 4×5
SUPER-ANGULON 90mm F6.8
f11 1/30 EKTACHROME

21   *Bamboos and cherry blossoms*
PENTAX 6×7
SMC TAKUMAR 300mm F4
f11 1/125 EKTACHROME

22   *Camellia red-bamboo green*
PENTAX 6×7
SMC MACRO-TAKUMAR 135mm F4
f5.6 1/8 EKTACHROME

23   *Bamboo roots and fallen camellias*
LINHOF SUPER TECHNIKA V4 4×5
SYMMAR 150mm F5.6
f22 1sec. EKTACHROME

24   *Volcano seen through a bamboo grove*
PENTAX 6×7
SMC TAKUMAR 105mm F2.4
f16 1/60 FUJICHROME

25   *Vine on bamboo*
PENTAX 6×7
SMC TAKUMAR 135mm F4
f5.6 1/5 EKTACHROME EPR

26   *Jikishi-an garden in Kyoto*
LINHOF SUPER TECHNIKA V4 4×5
SYMMAR 150mm F5.6
f11 1/60 EKTACHROME

27   *Designs of nature*
LINHOF SUPEA TECHNIKA V4 4×5
NIKKOR W210mm F5.6
f8 1/30 EKTACHROME

28   *Twisted bamboo sheath*
LINHOF SUPER TECHNIKA V4 4×5
TELE-ARTON 360mm F5.6
f11 1/60 FUJICHROME

29   *Shades of beauty*
LINHOF SUPER TECHNIKA V4 4×5
ZEISS-PLANAR 135mm F3.5
f11 1/30 EKTACHROME

30   *Stumps——aged and new*

LINHOF SUPER TECHROME V4 4×5
FUJINAR SC250mm F4.7
f22 1/2 EKTACHROME

31   *The red-tipped roots of a young bamboo*
LINHOF SUPER TECHNIKA V4 4×5
TELE-ARTON 360mm F5.6
f16 1/10 EKTACHROME

## Summer

33   *Fresh rain on a young bamboo*
LINHOF SUPER TECHNIKA V4 4×5
NIKKOR 210mm F5.6
f11 1/15 EKTACHROME

34   *The elegant shoots of "madake" bamboo*
PENTAX 6×7
SMC TAKUMAR 105mm F2.4
f8 1/125 FUJICHROME

35   *Crisp, pale beauty*
LINHOF SUPER TECHNIKA V4 4×5
FUJINAR SC250mm F4.7
f8 1/30 EKTACHROME

36   *Kyoto bamboo grove*
PENTAX 6×7
SMC TAKUMAR 105mm F2.4
f5.6 1/5 EKTACHROME

37   *On a rainy day*
LINHOF SUPER TECHNIKA V4 4×5
FUJINAR SC250mm F4.7
f11 1/8 FUJICHROME

38   *Weathering a typhoon*
⸱
39   PENTAX MX 35
SMC PENTAX M400mm F5.6
f5.6 1/60 EKTACHROME

40   *A sprinkling of yellow*
LINHOF SUPER TECHNIKA V4 4×5
SUPER-ANGULON 90mm F6.8
f16 1/60 KODACHROME

41   *Hillside bamboo grove*
LINHOF SUPER TECHNIKA V4 4×5
FUJINON SF250mm F5.6
f16 1/125 KODACHROME

95

96